MASTERING

SAMSUNG

GALAXY TAB S7 AND S7 PLUS

A Comprehesive User Manual with Step-by-Step Guide to Hidden Features of Samsung Galaxy Tab S7 and S7 Plus

KONRAD CHRISTOPHER

Copyright

Contents

CHAPTER ONE

THE EVOLUTION OF SAMSUNG GALAXY TAB SERIES

Ever since the very first model of the Galaxy Tab was introduced to the general public. The Galaxy Tab produced by Samsung to run on the Android platform, has been on an upward scale in its growth. Between the year 2010, when it first surfaced and now they have shown a lot of gradual advancement of this Mobile brand.

The 7-inch Samsung Galaxy Tab which was used to introduce the series to the general public at the IFA in Berlin sometimes in September 2010 and made available to the public on the 2nd day of November that same year, was the first Android powered Tablet to be released. It was developed to run on a Single Core 1Ghz Exynos 3110 processor and display with a resolution of 1024 X 600 pixel on a 7-inch TFT LCD. This device was the first of its kind and based on the Android 2.2 "Froyo" operating system. It also has a RAM space of 512mb or 444mb, internal storage of about 2GB and 16 to 32Gb microSD slot. The camera was a VGA located at the rear of the phone, and it operated over a removable battery of 1,500mAH, which were all big deals back then.

This version was immediately followed by the Galaxy Tab 7.0+ which, unlike the previous version, used a PLS panel and came pre-installed with an Android 3.2 HoneyComb version and also had some new hardware features on it but kept the initial screen resolution. Few months down the line, the Samsung Tab 8.9 was announced at the CTIA wireless convention. This device has a battery capacity of 6100mAH, an Nvidea Tegra 2 processor and a 3MP rear-facing camera. It weighs an approximate 470 grams, runs an Android 3.2 "Honeycomb" and has an update available to Android 4.0 "ice cream". It was finally made available on the 2nd October 2011. The phone continued growing with every new release and always seemed to consider user suggestions on what could have been done differently.

Here is a table of all Samsung Galaxy Tabs and the year they were released. (Table doesn't include S7 and S7+)

Galaxy Name	Year released
Galaxy Tab 7.0	2010
Galaxy Tab 7.0+	2011
Galaxy Tab 7.7	2011
Galaxy Tab 8.9	2011
Galaxy Tab 10.1	2011

Galaxy Tab 2	2012
Galaxy Tab 3	2013
Galaxy Tab 3 lite	2014
Galaxy Tab 4	2014
Galaxy Tab Pro	2014
Galaxy Tab S	2014
Galaxy Tab A	2015
Galaxy Tab E	2015
Galaxy Tab S2	2016
Galaxy TabPro S	2016
Galaxy Tab A 7.0 and 10.1	2016
Galaxy Tab S3	2017
Galaxy Tab S4	2018
Galaxy Tab S5e	2019
Galaxy Tab S6	2019
Galaxy Tab A 10.1	2019
Galaxy Tab A 8.0	2019

Samsung Tab S7 and S7+

These two tablets were both unveiled earlier this month, alongside the Galaxy Note 20.These devices seem like an answer to a lot of users' cravings. Unveiled in mystic black, mystic silver and mystic bronze colors, the phone's specifications are definitely attractive, as they, features perfect experiences

PC performance, connectivity and portability

Installed with a PC power capacity and a great hands-on portability, you can simply convert your tablet to a PC experience with the DEX and optional - the redesigned keyboard attachments - which now features the shortcut and function keys. It also features: Wi-Fi and a 5G connectivity option, for streaming, downloading and enjoying the internet at a very fast speed, without any viable lags, an optional expandable keyboard, a large 512GB storage space and a battery that can last workdays, a redesigned and better engineered S pen that can easily be used to navigate between windows, edge to edge screen and a processor that 30% faster than the previous S6 Tab.

Gaming and Movies

It also features a graphic packet that is a whole body of awesome to enable you game at the best way possible, with an immersive screen and Bluetooth enabled controller. This device is definitely not just a work thing; it offers as much features for entertainment as to work. With cinematic visuals that raises the bar so high and a big theatre sound, which uses a Dolby Atmos surround sound, and a Quad speaker that is tuned by AKG, you can watch more movies and enjoy them to the fullest with the edge to edge screen.

Body

It comes with a 285 x 185 x 5.7mm dimension and a weight of 575g; Glass front, Aluminium back and frame. It also features the following colors - Mystic Bronze, Mystic Black, and Mystic Silver.

Screen Resolutions

The Samsung S7+ Tab has a 12.4" AMOLED screen while the S7 comes with an 11" LCD TFT screen.

Camera and Battery

With a 13MP and an Ultra wide 5 Megapixel back facing cameras and the 8MP selfie camera you can capture the moments perfectly and have something to remember all the beautiful moments shared with your loved ones. The S7 and S7+ devices have a battery capacity of 8000mAH and 10,090mAH respectively; you can work/play with your device after just a single full charge for hours and then get back to 100% super-fast.

CHAPTER TWO

GETTING STARTED WITH THE SAMSUNG TAB S7 & S7+

It is important to note that the TAB S7 is the model version called SM T870, while the 5G version (TAB S7) is SM T875. TAB S7+ is the model version SM T976 (5G VERSION). SM T970 is the Wi-Fi only version of the TAB S7+

Now that your device has been shipped to your doorstep, the first thing is to unbox your Tab from the pack to see the accessories that come with your tablet. Inside the pack, you will see an adapter and a USB-C cable, which you will use to charge your device. You should not use your Tablet without charging it first.

Charging your Tablet

Kindly take note of the following before you start charging your device;

- Use only chargers approved for your device by Samsung for charging purposes. Using any other charger not approved by the manufacturer can cause charging problems.
- Save energy by unplugging your charger from the wall socket when not in use. The manufacturer does not incorporate a power switch in your charger, so unplugging the charger after use will prevent power wastage.
- While charging, your Tablet and the charger may get to hot; kindly carry out the following actions if you observed that your charger or the device has become hot during the charging process;
  - Remove the charger completely from your device (you can as well remove the charger from the socket if you can) and terminate all apps that might be running both in the foreground and the background. Exercise a little patience for the gadget to cool down before you start charging again.
  - If it is the bottom of your Tablet that gets overheated, it might be the issue with the USB-C

cable; kindly use another cable approved by the manufacturer to charge your gadget.

- o If you are using a wireless charger, avoid placing a metallic object between the wireless charger and the Tablet.

To charge your device with the charger, follow the steps below;

- Connect the USB Type C cable to the power adapter and plug the USB into your Tab's multipurpose jack.
- Plug the power adaptor into a secure wall socket that has been safely connected to electricity.
- Once your device is full, unplug the device from the charger and also detach the power adaptor from the wall socket (to avoid power wastage).

Charging tips for your battery

- The battery icon will bring empty once the device is low on power. Simply connect to the charger to start charging.
- It is not advisable to use the device until the power gets totally drained (power goes off your device); but if it gets totally drained, do not turn on the device just immediately after connecting the charger. Allow the gadget to charge for some few minutes before you power it on again.
- The battery will get drained quickly if you are running multiple applications, browsing with your device or sending data to another device. Endeavor to charge your

device fully before connecting your device to another device to avoid power loss.

- Using another source of power (like a computer) apart from compatible Samsung chargers may not be able to charge your device faster.
- There are many unfounded opinions on social media about not using your device while charging. This does not apply to your Samsung Tablet as you can very much use your gadget while you are charging it.
- If your Tablet gets an unstable power supply during charging, the touch screen may not work well. If this is the case, simply unplug the charger from your tablet.
- While charging your Tablet, the tab and the charger may become heated up. This is not uncommon and should not necessarily affect your gadget's lifespan or performance. If the battery becomes hotter than expected, the charger may stop charging. If this happens during wireless charging, disconnect the tablet from the charger to let it cool down, and then charge the tablet again later.
- Do not charge your tablet when the multipurpose jack is wet with water or anything liquid as the tablet might get spoilt. In case you notice traces of liquid in the multipurpose jack, kindly dry it first before you plug the device into the charger.
- If you notice that your tablet is not charging properly, take the gadget and its charger to a Samsung Service Centre.

Boosting your device's charging speed

The Samsung Tab S7/S7+ has inbuilt fast charging capability, and you can most times achieve a high charging speed if you turn off your gadget while charging. To enable the **fast charging** capability for your device, follow the step

- Go to the **settings app,** click on **device care** and select **battery** and then toggle on the **fast charging** switch to activate it.

Note that the fast charging capability can only be activated before you start charging, as you won't be able to activate this feature during the charging process. To deactivate the fast charging capability, make sure you are not charging at that moment, as you won't be able to deactivate it when you are still charging your gadget. When you are already charging your device with a standard battery charger, you cannot use the fast charging capability.

Installing SIM card on your device

You will only be able to use a Nano SIM card on your tablet. Get your SIM card from a recognized service provider. To insert the Nano SIM Card in your device, follow the steps below;

- Remove the ejection pin from the phone box (the SIM ejection pin accompanies your device).
- Locate the hole beside the SIM tray on your device and insert the ejection pin inside the small hole to pull out the SIM tray. You have to be careful so that you will not damage the SIM card and the tray.
- Put the SIM card on the SIM tray appropriately.

-   Carefully press the SIM card tray (with the SIM) into the SIM slot. Make sure the SIM tray is fully inserted into the SIM slot so that there won't be any entry for liquid to find its way into the SIM slot.

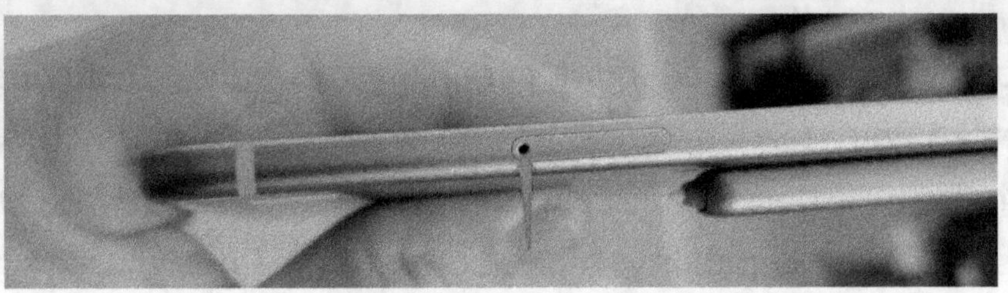

Installing a memory card on your device

Memory card capacity varies from device to device and not all memory cards will work on your device. Memory card compatibility depends on the manufacturer and the type of device that you are using. You don't use an incompatible memory card on your device as this can damage your tablet or the memory card, and can even implicate the files stored on the memory card. Your tablet works with the FAT and the exFAT file format for memory cards. When you want to use a memory card that has been formatted in another file format, your device will prompt you to reformat the SD card otherwise it will not be able to recognize it. Do not erase data you're your memory card every time as this shortens the memory card's lifespan. The memory card you put inside your tablet has its file directory in the **"My files → SD card folder."** To insert your memory card into your device, follow the steps below;

- Remove the ejection pin from the phone box. The SIM tray is also the memory card tray (the SIM ejection pin accompanies your device).
- Locate the hole beside the SIM & memory card tray on your device and insert the ejection pin inside the small hole to pull out the tray. You have to be careful so that you will not damage the memory card and the tray.
- Put the memory card on the tray appropriately.
- Carefully press the memory card tray (with the memory card) into the appropriate slot. Make sure the tray is fully inserted into the slot so that there won't be any entry for liquid to find its way into the SIM slot.

Removing memory card from your device

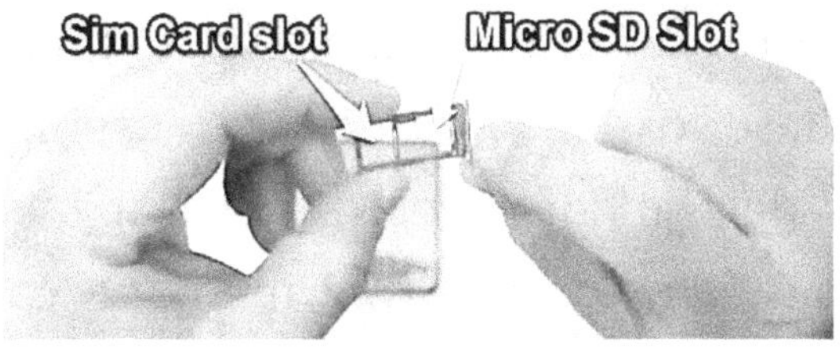

It is not advisable to remove the memory card from your tablet without first un-mounting the memory. To unmount memory card, follow the steps below;

- Navigate to the **settings app,** click on **Device care** and tap on **storage.**
- Select **"Advanced,"** tap on the **SD card** and click on **"Unmount."**

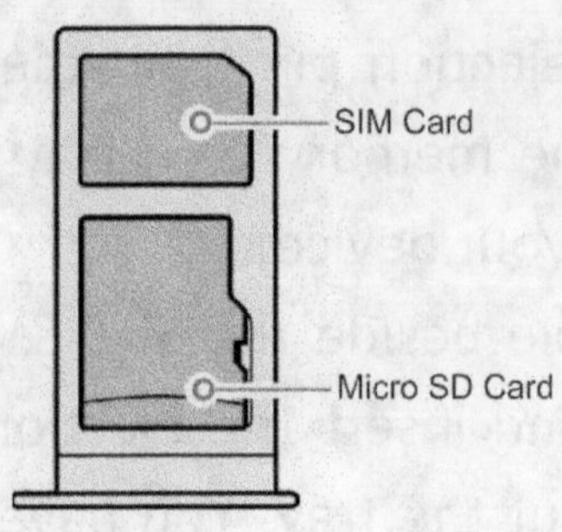

Tip: If you want to avoid unwarranted data loss or memory card getting spoilt, do not remove your memory card from its tray during file transfer. You might terminate the transfer process if you do so or damage your storage card.

Formatting your device's memory card

You need to transfer all useful information, files and contents stored on the memory card into another device or storage before you format the memory card. Failure to properly backup your content before formatting the memory card might cause loss of data or other useful information. Loss of users' data is not covered under any warranty from Samsung. You can format the memory card on your computer, but this is not advisable as most memory cards formatted on computers were later found to be incompatible with tablets and other devices. Kindly format your memory card right on your device. To format your memory card, follow the steps below;

- Navigate to the **settings app,** click on **Device care** and tap on **storage.**

- Select **"Advanced,"** tap on **SD card** and click on **"Format."**

Turning on/off your Samsung Galaxy tab

After you have successfully inserted your SIM card (for the 4G and 5G variant of the tab s7/s7+) and memory card, and then charged your device to full charge; the next thing, obviously, is powering on the device so that you can get started on the amazing Samsung experience.

- To **turn on** your device, look for the side key beside your device and then press and the side key for about 10 seconds. Since you are just turning on the device for the first time, you will be prompted to set up your device. Follow the instructions displayed on the on-screen to duly set up your device.
- You can **turn off** your device by pressing and holding on the side and volume down button at the same time. Alternatively, turning off your device can be done from the **notification panel** by simply swiping from the top of the panel and tapping on the **power off**⏻.
- **Force restart** your device by pressing on the **side key** and the **volume down** button simultaneously

You can customize the side key button for the purpose of turning off your device. This will leverage you to turn off your device whenever you touch and hold the tablet's side key. To do this, navigate to **settings app,** select **"Advanced**

feature", and tap on **"Side key."** Click on the **"power off menu"** which is accessible under **press and hold.**

The Samsung Account

The Samsung account leverages you to deploy a number of Samsung services provided by TVs, Samsung websites and mobile devices. If you want to access more information on the Samsung account, tap on **"settings app"** on the home screen, click on **"accounts and backup"** and tap on **accounts > Samsung account > more option ⋮ > help.**

Follow the steps below to create your Samsung's account;

- Open the **Settings app** and click on **Accounts and backup > Accounts > Add account > Samsung account**. You can alternatively open the **Settings app** and click on ⚪ .
- Tap on **Create account**.
- Follow all the instructions that will be prompted on the screen to create your Samsung account with ease.

Signing in to your Samsung account

Once you have successfully created your Samsung account, you can sign in to your account with your Samsung account ID and password. You can as well use your Google account for the sign-in purpose. To sign in to your Samsung account, follow the steps below;

- Open the **Settings app** and click on **Accounts and backup > Accounts > Add account > Samsung account**. You can alternatively open the **Settings app** and click on .
- Input your ID and password for your Samsung account and choose **sign in.** If you decide to sign in to your Samsung account with your Google information, you can select **"Continue with Google."**
- Follow all the instructions that will be prompted on the screen to log in to your Samsung account with ease.

Finding your ID and resetting your password

No one is an island, and we are liable to forget things. If you forget your Samsung's ID or you forget your Samsung's login password, don't beat yourself too hard. Simply click on the **"Find ID** or **Reset password"** on your sign in page. Input all the required information that you will be asked in order to reset your ID or Password.

Removing your Samsung account

Even though it is not advisable to remove your Samsung account from your device, anything can happen which might necessarily necessitate such. Note that once you remove your account from your tablet, you are automatically removing some of your data (such as events from calendar or contacts) saved on the account. To remove your Samsung account, follow the steps below;

- Open the **Settings app** and click on **Accounts and backup > Accounts.**

- Select **Samsung account**, tap on **personal info** follow by **more options** ⋮ and then tap on **"Sign out."**
- Once you choose **"Sign out,"** you will be asked to enter your Samsung password, input the password and then choose **"Ok."**

Transferring files or other contents from your old tablet or other devices to the new tablet

The Smart Switch

The Smart Switch feature affords you to seamlessly transfer your contents from the old device to your new tablet S7/S7+. To activate the Smart Switch, follow the steps below;

Requirement to use the Smart Switch

FOR PC

If you want to transfer contents from your PC to your new tablet with the Smart Switch feature, your device's specification must be of the following;

1) A Samsung tablet device (your new tab s7/s7+) that runs on Android OS version 4.3 or later.

2) Your old device that meets one of the specifications below; • your old Samsung tablet or other Android gadget running Android version 4.3 or advance • Apple product (iPhone or iPad Pro) running on iOS version 5 or bigger.

3) Your Windows platform featuring the following requirements: • Operating System: Windows 7, Windows 8 or Windows 10 • Central Processing Unit: Pentium 4; 2.4 GHz or bigger CPU • Random Access Memory: 512MB, 1GB, 2GB, 4GB or advance • Screen Resolution: 1024 x 768 (600), 32 bit or 64 bits • Software Required: Windows Media Player version 11 or advanced.

FOR MAC

If you want to transfer contents from your Mac to your new tablet with the Smart Switch feature, your device's specification must be of the following;

1) A Samsung tablet device (your new tab s7/s7+) that runs on Android OS version 4.3 or later.

2) Your old device that meets one of the specifications below; • your old Samsung tablet or other Android gadget running Android version 4.3 or advance • Apple product (iPhone or iPad Pro) running on iOS version 5 or bigger.

3) A Mac platform featuring the following minimum requirements: Operating System: Mac OS X 10.9 or bigger version, Central Processing Unit: Intel Core 2 Duo 2.0 GHz or bigger version, Random Access Memory: 512MB, 1GB, 2GB, 4GB or advance, Screen resolution: 1280 x 800, An application that can transfer files on Android installable on your Mac.

FOR MOBILE APP

For Android phones: - Wireless transfer requires the following: Android OS starting from 4.3 or advance. Wireless transfers from an Android device that is compatible to your Galaxy Samsung Tab S7/S7+: Android OS starting from 4.3 or more (if you have a device with OS below 6.0 that is not Samsung, you will only be able to connect with Samsung device that has a mobile APP). - Wired transfer: Android 4.3 or bigger version, a USB connector and a cable (the one you use for charging).

For devices running on iOS – choose any one of the two listed alternatives that better work for you: - Wired transfer from your iOS gadget to your Galaxy Samsung Tab S7: requires OS starting from iOS 5.0 or bigger version, a USB connector and an iOS device cable (lightning or 30 pin). - Importing from iCloud requires: iOS 5 or bigger version and Apple ID - PC/Mac transfer using iTune requires: Smart Switch for PC/Mac software (download from Samsung web).

BlackBerry gadgets: - Wireless transfer requires BlackBerry OS 7 or 10 (Mobile AP) - Wired transfer from your BlackBerry device to your Samsung galaxy tab requires: BlackBerry OS 7 or 10 for transfers by using your USB connector

Windows Mobile phones - Wireless transfer requires: Windows OS 8.1 or 10.

Transferring contents from your old Android phone to your new Samsung Galaxy Tab S7/S7+

A wireless transfer is a very good way of copying your information or contents from your old Android phone to a new gadget (Samsung Galaxy Tab S7/S7+ in this case). It offers a very fast and secured means of transferring content while also making sure both devices are plugged in for charging; since you are not making use of their charging ports for transfer.

- To begin with, you are required to make sure that the **Smart Switch** app has been successfully installed on the old device. Your Samsung Galaxy tab S7/S7+ already have the Smart Switch feature by default from the manufacturer. You can open the Smart Switch feature on the new Tab S7/S7+ by clicking on the **"Settings app,"** and then click on **"Accounts and backup"** and tap on the **"Smart Switch"** feature. Place the Samsung Tab S7/S7+ about 4 to 5 inches away from your old Android gadget to get started.
- Since you have toggle on the **Smart switch** feature on your new Samsung Tab, simply go to your old Android device and open the **Smart Switch** app. If you don't have the Smart Switch app on your old phone, do not panic. Simply go to the galaxy store (if it is a Samsung device) or to the app store (for other android devices) to download the **Smart Switch** app for your old device. Choose **Send data** from the Smart Switch on your old Android phone, choose **Receive data** on the new phone (Samsung Galaxy

Tab), and then choose **Wireless** on both gadgets. On the new Samsung Tab device, you may be instructed to select the type of your previous (old) phone. Follow all the instructions which will be prompted for you on the screen.

- Now, activate the connection between the two devices by tapping on **Allow** on the old device. On your old device, tap all the content you want to send, and then click on **Transfer**. Once the transfer has been completed, tap on "**Done**" on your new device and that should be all.

Transferring your files from your old iPhone to your new Samsung Galaxy Tab S7/S7+

Toggle on the **Smart Switch mode** on your new Samsung Galaxy Tab device by clicking on the **"Settings app,"** and then click on "**Accounts and backup**" and tap on the "**Smart Switch**" feature. Tap on **Receive data** from your new Samsung Galaxy Tab. Select **Wireless**, and then choose **iPhone/iPad**. Input your Apple Identification (Apple ID) and password, and then tap on **Sign in**. Fill in the verification code, and then tap on **OK**. Select the file/ contents you want to move, choose **Import**, and then tap on **Import** once again.

Note that you will not be able to transfer iTunes music and videos from your iCloud. You can copy the M4A files (for iTunes music that are un-encrypted) from the PC through your iTunes library. Transferring your music files to your new Samsung Galaxy Tab by sending such music first from your iPhone to the PC using a transfer cable is also possible, you can then send from PC to your new Samsung Galaxy Tab with a transfer cable. iTunes videos and music cannot be transferred from iCloud. iTunes music that is essentially unencrypted can be transferred to your phone by literally copying the M4A files from a PC using your iTunes library. You will also be able to transfer your music files directly from your PC.

Transferring your files using your USB cable

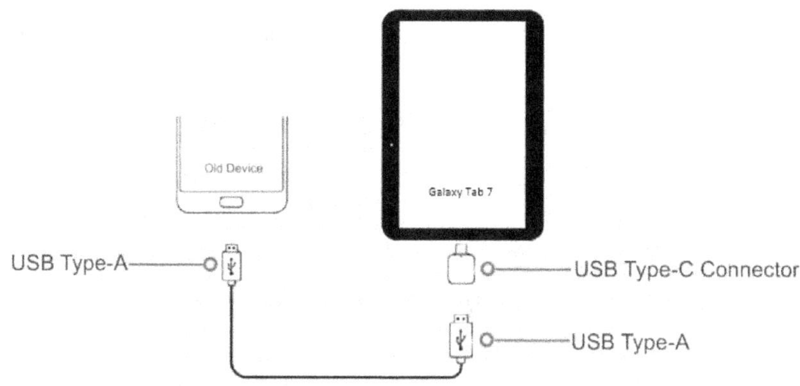

A wired transfer is possible with your USB cable. This is preferable mostly when you don't have too many files to transfer from your old device to the new device (Samsung Galaxy Tab). During a wired transfer, you will not be able to charge your Samsung Galaxy Tab S7; this is because the charging port is busy as it is the one you are using for the transfer. Hence, you need to ensure that your device has been fully charged if there is any need for the transfer to take much time.

Get your USB-OTG adapter that you can use to connect both phones. To get started, simply connect your new Samsung Galaxy Tab to your old device's USB cable. Most cables will require you to deploy a USB-OTG adapter. But if you have a USB-C to USB-C cable or a Lightning to USB-C cable, you won't be needing any adapter, as the USB-C will fit right inside the new Galaxy Tab.

1. Open the **Smart Switch** on both devices.
2. Click on **Send data** on your old device, choose **Receive data** on your new device (Galaxy Tab), and then choose **Cable** on both devices. Smart Switch will scan your old device for any file or information that can be transferred.
3. Mark all the files or information that you want to send to your new Galaxy tab. A total value of the transfer time will be displayed on the screen. If the whole transfer will use more than 1 hour, you can use a wireless transfer instead so that both gadgets can be charged during the process.
4. Tap on "**Transfer**" to start sending your files or contents.
5. When the transfer has been done successfully, select "**Done**" on the new device, and click on "**Close**" on your former device.

Transfer files using external storage

If you have in your possession a sizable microSD card or a USB OTG flash drive, transferring your files or contents from your old phone to the new one by using the Smart Switch is a walk in the park. But before you get started on this amazing journey, you will need to put your SD card or the flash drive inside your old device.

Open the **Smart Switch** app on your former device. Tap on the **SD card** icon, (you can see the SD icon at the upper right corner of your device's screen) and then tap on **Back up**. Mark all the files that you want to back up. If your SD card doesn't have enough space, you can unmark some selections and later come back to select and send those selections you have unmarked. Tap on **Back up** once again. When the transfer has been done successfully, click on "**Done**."

Next, insert the external storage device into your new Galaxy Samsung Tab device, and then open the **Smart Switch** feature. Tap on the **SD card** icon, and then select **Restore**. Mark all the items or contents that you would like to restore, and then tap on "**Restore**" once again. When you are done, click "**Done**."

Transferring files from your Windows and Blackberry phones

To transfer contents from your Windows to the new Samsung Galaxy Tab, click on "**Receive data**" On the new Tab and choose **Wireless**, then tap on **Windows Phone/ Blackberry**. The complete steps you can follow to download the Smart Switch application on your old device (Blackberry or Windows) will be brought to you on your old gadget.

Follow strictly all the instructions on the screen to get started. As soon as you are done, select **"Connect"** on your old device and then choose the network. Type the password shown on your new Samsung Galaxy Tab and your content will now be transferred successfully.

To access the Smart switch on your device;

- Select **"Accounts and backup"** from **"Settings"** and choose **"Smart Switch."**

Follow all the procedures above to get started with transferring your files depending on which device you are using and the device that you want to transfer to.

The Navigation bar (contains the navigation button)

After you have turned on your device's screen, you will see three soft navigation buttons at the bottom of the screen. As their name implies, these three soft buttons are used majorly for navigation purposes. You can switch between screens, navigate to the home screen and carry out some other major tasks with the help of these three navigation buttons. Their functions can actually vary based on the application you are currently using and the environment in which you are using the navigation buttons and the apps. The three navigation buttons are as follow;

- **|||** **Recent:** Quickly tap on the **recent** navigation button to access the list of apps you have recently opened.

- ◯ **Home:** This navigation button will take you to the home screen wherever you are on your device. You can also open the **Google assistant application** by tapping and holding on the **home button** for about three (3) seconds.
- ‹ **Back:** You will be returned to the previous screen when you tap on the back button.

If you want to enjoy viewing your applications or systems on a much wider screen, you can hide the navigation bar. To hide the navigation bar, follow the instructions below;

Hiding the navigation bar from view

- From the Settings app, click on **Display**, select **Navigation bar**, and then choose **Full screen gestures** located beneath the **Navigation type**. When you do this, you have successfully hid the navigation bar and you will now see the gesture hint.
- Tap on the "**More options**" and choose a desired from the list below;
  - Swipe from bottom: The gesture hints will show up where the soft buttons are situated. To use a particular soft button, simply drag the gesture hint of that particular button upwards.
  - Swipe from sides and bottom:
    - You can direct yourself to the home screen by simply swiping up from the bottom of your device's screen to move to your Home screen.

- Swipe up and then hold to see the list of apps you have opened recently.
- You can return to your previous screen by swiping inward from either screen's side.
- Open the **Google Assistant** by swiping inward from one of the corners at the bottom.

You can hide the gesture hint at the bottom of the screen by toggling the **gesture hints** switch to deactivate the gesture hint.

The Home screen and the App screen

The Home screen provides you with full access to your device's functionality and features. It features the widgets, app shortcuts, navigation bars and some other things. The App screens show all the apps that you have installed on your device, including the apps, you recently downloaded. The App screen is found on the home screen.

You can navigate between the home and the app screen by swiping upward or downward from the home screen to have access to the applications screen. You can get back to the home screen from the app screen by swiping upward or downward on the application screen. You can alternatively use the home button or the back button to switch between the home screen and the app screen.

Searching contents on the home screen using the app finder

- From the app screen, click on **app finder**. You can alternatively open the notification panel, swipe downward from the notification panel and then click on Q
- Input a keyword to search whatever content you want to search.

Moving items

You can move an app on the app screen to another location by tapping and holding on such an app and then drag it to any location that you want. You can also move an item to a new panel by dragging the app to the screen's side.

Tap and hold on any app on your app screen to add a shortcut to the app on the Home screen. You can as well drag frequently used apps to the shortcuts area at the bottom of your device's Home screen.

Creating folders for apps on the Home screen

Similar apps on the home screen can be grouped together into a folder for easy accessibility. Touch and hold on the app, and then drag the app over another app. A new folder will be created that will contain the application you recently dragged. Click **folder name** and input the name you want for the folder.

Adding more apps to the folder

- Click on the **add icon** ✚ on the folder. Mark all the apps that you want to add inside the folder and choose **"Done."** You can as well add any app into the folder by dragging the app into the folder.

Moving apps from a folder

Tap and hold on to any app that you want to move and then drag the app to a new location.

Deleting a folder

Tap and hold on to any folder that you wish to delete, and then select the **Delete folder.** Deleting an app folder will not delete the applications inside it, only the folder will be removed. The applications inside the folder will be moved to the Apps screen.

Editing the Home screen

On your device's Home screen, tap and place your fingers over an empty area, or bring your fingers close together to bring the editing options. You will be able to add widgets, set the wallpaper, and do much more. You will also be able to add a home screen panel, delete a home screen panel, or rearrange Home screen panels.

- **Adding panels:** Swipe to the left to add panel, and then click on ⊕.

- **Moving panels:** Tap and place your fingers over a panel preview, and then move it to another location.
- **Deleting panels:** Tap on the delete 🗑 icon to delete a panel.
- **Wallpapers:** Change the settings for the wallpaper for your Home screen and the locked screen.
- **Widgets:** Widgets are small applications that initiate specific app functions to give information and easy access on your Home screen. Tap and place your fingers over a widget, and then drag the widget to your home screen. The widget will be included on the Home screen.
- **Home screen settings:** Configure various settings for your Home screen, such as the layout or screen grid.

Displaying all apps on the Home screen

If you don't want to use a separate apps screen, you can use your tablet to show all applications on your home screen. On your home screen, tap on an empty area and hold on it, and then click on **Home screen settings**. Select **Home screen layout**, tap on **Home screen only** and choose **Apply**. You can now see all your applications by swiping to the left on your Home screen.

CHAPTER THREE

IMPORTANT ICONS IN SAMSUNG GALAXY S7 AND S7+ AND THEIR MEANINGS

The Indicator Icon

You can locate the indicator icon on your device's status bar at the top area of the screen. Note that when you are using some applications, you may not be able to see the status bar because these apps have already occupied the whole screen. You can swipe down from the top of the screen to bring the status bar. Some of the indicator icons will only appear when you open the notification panel. The listed indicator Icons can be found on your device;

Network Icons

- ⊘ **No signal:** This tells you that your device service connection is out of range. It may be difficult to make and receive calls when you see this icon on the indicator icon.

- ▫ **Signal strength:** This indicates the strength of the signal in your locality. When the bar is full, it means your device's signal is excellent and you can send and receive calls without any interruptions.

- ▫ **Roaming:** Outside of normal service range. With this, you will be able to send and receive voice calls, send and receive data or navigate some other services when you travel out of your geographical area using a visited network. If you move out of the United States to the UK, you will be using a roaming service and you will be able to have access to the internet when at the price of a local call or at prices less than the regular long distance charges.

- **G GPRS Network connected:** Means the general packet radio service on the 2G and the 3G cellular network that gives you moderate speed data transfer using unused time division multiple access (TDMA) channels.

- **E Edge Network connected:** This is the enhanced version of the GPRS network and is used on top of GSM networks. This network gives you almost three times faster speeds than the normal and outdated GPRS.

- **3G UMTS Network connected** Meaning Universal Mobile Telecommunications system. Gives you better spectra efficiency and bandwidth.

- **H HSDPA network connected:** Meaning High-speed downlink packet access. This is an enhanced version of 3G in the high-speed packet access family. It enables networks that are based on UMTS to have higher data speed and efficiency.

- **H+ HSPA+ Network connected:** Also referred to as 4G or the evolved HSPA.

- **LTE Network connected:** Meaning Long-term evolution. Standard for wireless transmission that leverages you to download contents or files online at some very high speeds. Faster than the 3G.

- **5G Network connected:** Ultra-fast speed. It is a 5th generation network and it is currently being supported in the USA, South Korea and China.

- **Wi-Fi connected:** This icon will be shown when you are currently connected to the internet using a Wi-Fi network.

- **Bluetooth feature activated:** Tap on this icon to send files or contents that are not especially bigger in file size.

S Pen Icons

- **S Pen connected:** This is shown when you have connected your S Pen.

- **S Pen disconnected:** You see this when you have disconnected the S Pen.

- **S Pen battery power level:** This indicates the current power level of your S Pen. If it is low, kindly charge the S Pen before you start using it.

Location services being used currently: When you see this at the top of the status bar, it means that the app you are working with has turned on your location service.

Calls icons

- ✆ **Call in progress:** You will see this icon when you swipe down the notification panel. It tells you that you are currently on call. Any other call that enters when this icon is still active will enter as a missed call and you can always see the missed call notifications when you swipe down on the notification bar.

- ✆ **Missed call:** This icon will be displayed at the top of the status bar and at the bottom of the notification panel. Kindly swipe up from the notification panel to see the number of calls you have missed when you are bot around and then attend to those calls.

Battery icons

- 🔋 **Battery charging:** You will see this icon at the top of the status bar and it shows that you have connected the right charger to your device and your tablet is currently being charged. Kindly remove the charger and your device once it is fully charged.

- 🔋 **Battery power level:** This indicates the current battery volume or power carrying capacity or percentage of your battery. Once the volume has come down tremendously, kindly connect your charger to start charging your device.

⚠ **Error occurred or caution required:** One of your applications or phone systems needs attention. Kindly swipe down from the top of the status to check which application actually needs attention.

📭 **New Texts or multimedia message:** Swipe down from the top of the status bar to see new messages. Tap on the message icon if you want to send a reply for the message.

📵 **Mute mode activated:** Your device is currently being muted

✈ **Flight mode activated:** You won't be able to access the internet, send or receive calls and messages. Activate this icon when you want to be left alone for sometimes. The flight mode is also a safe option when you are travelling, as they will always advise you to put your device on flight mode.

📳 **Vibration mode activated:** This will put your device only in vibration mode. When a call or a message enters your device, your device will only vibrate so that you can be notified of the call or message.

⏰ **Alarm activated:** Use this icon to set an alarm.

Lock screen

When you hit the side key on your device, you are literally turning off the screen and unlocking it. Your device's screen will also turn off automatically and lock itself after a few minutes of inactivity. If you want to unlock your screen, swipe to any direction when the screen brings on light (turns on). If your screen is off, kindly hit the side key to turn it on.

You can alternatively tap twice on the screen to unlock the screen.

Changing the ways you lock your screen lock

If you wish to change the way you lock your screen, open the **settings app,** click on **lock screen > screen lock type** and then choose a method. Locking your device's screen prevents unauthorized access to your device. If they don't have your password, PIN, or your biometric information, there is no way they can access your device without your consent. Once you have chosen a lock screen method, your device will prompt you to input the code before you will be able to unlock your device. The following methods can be use;

• **Swipe:** Swipe on your device's screen in any direction to unlock your device's screen.

• **Pattern:** Unlock your screen by drawing a pattern with four or more dots to unlock the screen.

• **PIN:** Input a PIN with at least four (4) numbers to unlock your tablet.

Password: Input a mixture of characters, symbols or numbers to unlock your screen. The password must be at least four characters, numbers or symbols.

• **None:** Do not use a screen lock method at all

- **Face:** Use your face to unlock your tablet by registering your face appropriately.

Fingerprints: Register your fingers on your device's fingerprint scanner to unlock your tablet.

You can make your tablet perform a factory data reset once you input incorrect codes many times in a row and reach the maximum limit. To do this, open your device's **settings app** from the home screen, click on **lock screen** and choose **secure lock settings.** Unlock your screen by using the preset screen lock method, and then toggle the **"Auto factory reset"** switch to switch it on.

Notification panel

When you get new notifications, the indicator icons show on your device's status bar. To access more details about the icons, open the notification panel and check the details. To open your notification panel, swipe the status bar downward. Swipe the notification upward on the screen to close the notification panel. The functions below can be deployed on the notification panel. To close the notification panel, swipe upwards on the screen. You can use the following functions on the notification panel.

Using quick setting buttons

Click on the quick setting buttons to set some certain features. To see more buttons on the notification bar, simply swipe downward on the notification panel. To change feature settings, click on the text under each button. To access more detailed settings, click and hold a button.

You can rearrange buttons by tapping on the **more option** icon, choose **Button order**, click and hold a button, and then move the button to another favorite location.

Controlling media playback

You can have access to play your favorite music or watch videos with the media menu without necessarily opening their applications from the home screen. All that you need to do is to take control of the media playback feature. You can as well continue to play the music or the video on any other device. To do this, follow the steps below;

1. From your notification panel, select **media.**

2. Click the icons on the controller if you want to control the playback.

To continue the playback on another device, click on ⅇ and chose a device you want.

Controlling nearby devices

Quickly launch quickly and take adequate control of connected devices that are nearby your tablet and occasionally used connected devices and frequently used SmartThings gadgets and scenes on your notification panel.

- ✓ From the home screen, swipe down from the top section of your device to have access to the notification panel, and then click on **Devices.**
- ✓ Choose a nearby device or a nearby SmartThings gadget to start controlling it, or choose a scene to launch the scene.

Screen capture and screen record

Screen capture

Use your device to capture any screenshot on your screen. You can be surfing the internet and come across any content that catches your attention, simply use the screenshot feature to take the page and save it. You will be able to do some editing, such as writing on the screenshot, draw on the screenshot, crop and share the screenshot.

Your screenshots are usually saved in the Gallery application and can be accessed from the gallery. The following methods can be used to capture your screen;

- **Key capture:** Use your device's side key and the volume down key to capture the screen by pressing both keys simultaneously.
- **Swipe capture:** Capture your current screen by swiping your hand to either the right or left across your device's screen.
- **Air command capture:** Use your S Pen to take a screenshot of your current screen. To do this, simply hover the S Pen on the screen and then hit the S Pen

button to have access to the Air command panel, and then, tap on **Screen write**. More on the S Pen later.

- o Some apps and features do not permit taking screenshots of the current page.
- o If you cannot take a screenshot by swiping, that shows the feature has not been activated. Kindly activate by going to the **settings app,** click on **'Advanced features,"** select "**Motions and gestures,"** and then capturing a screenshot by swiping is not activated, launch the Settings app, tap Advanced features $\rightarrow$ Motions and gestures, and then toggle the "**Palm swipe to capture"** switch to enable it.

After taking a screenshot, utilize the options below on the toolbar at the bottom of the screen:

- ❖ To capture your current content and the hidden content on an extended page, like a webpage, simply tap on . When you tap the icon, the screen will automatically scroll down and more screenshots will be captured.
- ❖ Allows you to draw on the screenshot or to crop a part from the screenshot. You can access the cropped part in the Gallery app.
- ❖ # This allows you to add tags to your screenshot. You can search for your screenshot by tag when you tap on the **finder** located at the top section of the application screen. You will be able to access the tag list and quickly navigate any screenshot you want to search.
- ❖ Share the screenshot with others. Tap on this icon to share your screenshot with anyone. You will be asked to

select the sharing medium, simply select the medium that you wish to share the screenshot to.

If you cannot see the four options above on the captured screen, quickly open the **settings app,** select **"Advanced features > Screenshots > Screen recorder"** and then click on the screenshot toolbar switch to enable the toolbar.

Screen record

You can record all that your screen is doing while still using your device. When you give your device's access to anyone, you can secretly activate the screen record to keep a record of all the features they are using on your phone, and you can then watch it later. To do this, follow the steps below;

- From the home screen, open your notification panel, swipe downwards from the notification panel, and then choose **Screen recorder** to enable the screen recorder.
- Choose a sound setting and then hit **Start recording**.

After a few seconds, your recording will start.

- You will be draw or write anything on the screen by tapping on .
- Use a video overlay of yourself to record the screen by tapping on when you are done recording your video, click .

Your screen record video is accessible in the Gallery app.

To change the screen recorder settings, open the **settings app,** select **"Advanced features > Screenshots > Screen recorder"** and then tap on **"Screen recorder settings."**

Entering text

Keyboard layout

You will see a keyboard appearing automatically whenever you enter text to create notes, send emails and do something related to messages. Some languages are not supported in text entry on mobile. You need to change the language to any supported language.

Changing the input language

To change the input language, click on **settings**⚙, choose **Languages and types,** select **Manage input languages** and choose the languages that you want to use. When you choose two or more languages, you will be able to switch between the entry languages by either swiping to the right or to the left on the space key.

Changing the keyboard

On the navigation bar, click ⌨ to change your keyboard. To change your keyboard type, tap⚙, click on **Languages and types**, choose a language, and then select the type of keyboard that you want.

If you cannot see the keyboard button ⌨ on the navigation bar, open the

Settings app, click on **General management**, choose **Language and input**, select **On-screen keyboard**, and then hit the **Show Keyboard button** switch to enable it.

Additional keyboard functions

Some of the features below may not be seen on your keyboard depending on the service provider or your region.

• ☺ : Use emoticons in your texts.

• 😊 : Use stickers in your texts. You can as well input My Emoji stickers that resemble you.

• GIF : Use animated GIFs.

• Tσ : Use handwriting to enter texts.

• ⌨ : Change your keyboard mode.

• 🎤 : Tap on this to input text by voice.

• ⚙ : Change the settings for the keyboard.

• ••• → 🔍 : Search for content and input it.

• ••• → 🗚 : Translate your text and enter it.

- ▪▪▪ → (Pass) : Input your personal details registered to Samsung Pass.

- ▪▪▪ → ▶ : Share video links.

- ▪▪▪ → 🖫 : Insert an item from the clipboard.

- ▪▪▪ → ⌄ : Gain access to the text editing panel.

- ▪▪▪ → ▢ : Change the size of the keyboard.

- ▪▪▪ → 🙂 / 😃 / 🖼 : Use stickers

Copying and pasting

✓ To copy a text, tap the text and hold over it.
✓ Drag the ◖ or the ◗ to highlight the desired text. You can tap "**Select all**" to highlight all text.
✓ Once you have marked your desired texts, select **Copy** to copy the texts or **Cut** to delete the texts.
The selected text will be copied to the clipboard.

- To paste the text you copied, Tap and hold on the area where you want to paste the text to and then choose **paste.**
If you want to paste a text that you have long copied, click clipboard and then select the text.

Dictionary

Look up the definitions for some words while you are using a specific feature, such as when you are browsing a web page.

1. Tap and then hold over the word that you wish to look up.
2. Select Dictionary from the list of options. If a dictionary has not been pre-installed on your tablet, click on "**Move to Manage dictionaries**," tap the download icon next to a dictionary, and then choose **Install** to download it.
3. Check the definition of the word in the pop-up window of your dictionary.

To get the full screen view, click on . Click the definition on your device's screen to see more definitions. In the detailed view, click on ★ if you want to add such word to your list of favorite words or tap on the **Search Web** to utilize the word as a search term.

CHAPTER FOUR

APPS AND FEATURES

Installing or uninstalling apps

Galaxy Store

You will be able to purchase and download applications from the Galaxy store, particularly those applications that are made especially for Samsung devices.

To download and install applications from the Galaxy store, you need to open the Galaxy Store app on your tablet. The Galaxy store app's availability depends on your region or the service provider.

Installing apps

You can browse apps by category or click on the finder^Q to look for a keyword. Tap on an app to get more information about the app. To get free apps, click Install. Some applications are not available to be downloaded freely, that means that you will have to pay a certain charge to buy the app before you can install the application. To purchase and download an app where charges are applicable, click on the price and then follow the on-screen instructions to get the app.

Click on≡, select **settings** , tap on **"Auto updates apps"** and then select an option.

Play Store

You can download some applications freely from the play store, while some apps can only be downloaded when you purchase them. To buy and download an application, click open the play store from your home screen.

Installing apps

You can browse apps by category or click on the finder^Q to look for a keyword. Tap on an app to get more information about the app. To get free apps, click Install. Some applications are not available to be downloaded freely, that means that you will have to pay a certain charge to buy the app before you can install the application. To purchase and download an app where charges are applicable, click on the price and then follow the on-screen instructions to get the app. Click on≡, select **settings** , tap on **"Auto updates apps"** and then select an option.

Managing apps

Uninstalling or disabling apps

To uninstall or disable an application, simply tap and hold on an application and then choose an option.

• **Uninstall:** When you hold an app and tap uninstall, you will be able to uninstall the downloaded app. Note that if you uninstall an app, you will no longer be able to use such an app until you reinstall it.

• **Disable:** There are some default applications (apps that came pre-installed on your tablet) that cannot be uninstalled, simply disable these apps if you don't want them again. There are some applications that don't support being disabled.

Enabling apps

To enable an app, simply open the **Settings app**, select **Apps,** tap on ▼ and then click on **Disabled**, choose an app, and then select **Enable**.

Setting app permissions

Some apps will not work properly if they cannot access or deploy some information on your device. You need to give permission to these apps to be able to access information on your device.

To see your app permission settings, open the **Settings app** and click **Apps.** Choose an app and click **Permissions**. You can access the list of permissions that are available for the app and then change the permission you want to change. If you want to access or change permission setting for an app, open the **settings app** on your device, select **app,** click on the more options icon ⋮ and tap the **Permission manager**.

Choose an item and select an application. If you fail to grant permission for an app, some app's basic features may not work as they should.

S Pen features

Air actions

There is an S Pen connected to your tablet via Bluetooth Low Energy (BLE), which you can use to control most apps remotely.

For instance, you can urgently open apps, such as your camera app, by simply pressing and holding your tab's S Pen button. Also, when you are using your camera app, you can easily take a picture by pressing the S Pen's button once. While playing your favorite music, you can increase the volume by lifting up the S Pen while simultaneously pressing and holding your S Pen's button, and at the same time lower the volume by lowering the S Pen button. To activate this feature, open your **settings app,** click on **"Advanced feature"**, select **"S Pen"**, tap on **"Air action"** and then toggle on the switch to enable this feature. Note the following tips while using S Pen;

- You are expected to charge the S Pen before you start using the air action feature.
- When you remove your S Pen from the S Pen charger, you will see this icon on your status bar. If you discharge

your S Pen from its slot and take it far away from your device, or if there are obstacles or there exists an external blockage between your S Pen and the tablet, your S Pen will lose connection with the tablet and the S Pen icon will look grey . To reaffirm connection for the S Pen, simply connect it back to your tablet by putting the S Pen back in its slot.

Taking pictures with the S Pen

You can quickly take pictures by pressing your S Pen button. Although, your device's camera can always take pictures by setting your camera at a distance and setting the timer; but the S Pen gives a quicker way. To use your S Pen to take pictures, follow the steps below;

- ✓ Open your Camera app.
- ✓ Tap your S Pen button once to snap a picture.
  - o Change the Camera shooting mode by hovering your S Pen to the right or to the left while simultaneously tapping and holding your S Pen button.
  - o Tap your S Pen button two times to switch between cameras.

Changing apps or features to use

You can change the feature, app, or actions that you want to activate with your S Pen feature. If you want to change the app you want to launch with the S Pen, launch the Air actions setting screen, select "**Hold down Pen button to**," and then choose an app or feature. To change actions for each app, open the Air actions setting screen and select an app under App actions. You can change actions by tapping items.

Connecting another S Pen

If you want to use another S Pen with your device's S Pen features, like the air action, you have to connect the new S Pen with your tablet first.

- Attach the new S Pen to your S Pen charger.
- Launch the notification panel, swipe downwards, click S Pen air actions, and then click again to reactivate the S Pen.
- Click on "**Connect new S Pen.**"

Your S Pen will then be connected for use. You may be delayed a bit before the S Pen connects in the long run, kindly exercise a little patience. Note the following about connecting S Pen;

✓ You will only be able to connect Samsung-approved S Pens that support BLE (Bluetooth Low Energy).
✓ Do not remove your S Pen from your tablet once it has been connected. If you detach the S Pen, the conn-ection will be lost.
✓ If your S Pen refuses to connect or you just want to deploy the former S Pen, kindly reconnect the S Pen with the steps shown above.

Resetting the S Pen

If your S Pen has connection issues or the S Pen disconnects quite often, kindly reset your S Pen and then reconnect it one more time. Attach your S Pen to its charger. Then, launch your Air action setting screen and click the **more option** icon and then tap "**Reset S Pen**."

Air command

Air command is the menu that gives S Pen features and offers quick access to applications that are most frequently used.

To launch the Air command panel, move your S Pen over your device's screen and then tap the S Pen button. You can also use your S Pen to click on the Air command icon.

To choose the application or feature that you want, simply drag upwards or downwards on the Air command panel. When you have locked your device with a password or any other means, or your device's screen is off, removing your S Pen will not open the air command panel.

Adding shortcuts to the Air command panel

You can add shortcuts to apps or functions you frequently used functions or apps to the Air command panel. On the Air command panel, select "**Add shortcuts**" and choose apps or functions you want to be opening from the panel. You can alternatively add shortcuts to apps or functions by opening the Air command panel and then tap **settings** ⚙ → **Shortcuts.**

S Pen features

You can use the air command panel to enable some S Pen's features. Add any shortcut that does not show on the panel by default using the **"add shortcut"** option. Some of the S Pen features are;

• **Create note:** Write notes with ease in a pop-up window without necessarily opening the Samsung Notes application.

• **View all notes:** Launch your Samsung Notes application and see a listing of all the notes you have written.

• **Smart select:** Use your S Pen to highlight an area and carry out actions, such as saving or sharing.

• **Screen write:** Capture the screenshots you took and then write or draw on the screenshot. You can as well capture the current content and the hidden content on an extended page, like a webpage.

• **Live messages:** If you don't want to use the text message feature, you can create and then send a special message just by recording your actions while handwriting or drawing a live message and then saving the live message as an animated file.

- **AR Doodle:** Make some fun videos featuring virtual handwriting or some drawings on faces.

- **Translate:** Hover your S Pen over a particular to have the word translated.

- **PENUP:** Post your drawings, see others' artwork, and benefit from useful drawing tips.

- **Bixby Vision:** Deploy the Bixby Vision features in searching for similar images, detect text and then translate the text and a lot more.

- **Magnify:** Hover your S Pen over any area of your device's screen to enlarge it.

- **Glance:** Reduce any application to a thumbnail and move your S Pen over the thumbnail to launch the application in full screen view.

- **Coloring:** Give your images some color tints by using the S Pen.
- **Write on calendar:** Open your Calendar app to draw or write on the screen.
- **Add shortcuts:** You can add shortcuts to apps or functions you frequently used functions or apps to the Air command panel. On the Air command panel, select "**Add shortcuts**" and choose apps or functions you want to be opening from the panel. You can alternatively add shortcuts to apps or functions by opening the Air command panel and then tap **settings** ⚙ → **Shortcuts.**

Air command icon

You will still see the air command icon on the screen after the air command panel has been closed. You can launch the air command panel by tapping on the air command icon using your S Pen. When the Air command panel is closed, the Air command icon will remain on your screen. You can open the Air command panel by tapping the icon with the S Pen.

- To move the Air command icon, drag the icon to a new location.

- To remove the Air command icon, drag the icon to "**Remove**" which you can find at the top of the screen.

If the Air command icon is not showing on the screen, open the Settings app, click on "**Advanced features,**" **tap on** S Pen, and then click the "**Show floating icon**" switch to enable it.

You will be able to see actions that are available with each application while using applications that support the Air action feature. To do this, simply hover your S Pen over the Air command icon.

Create note

Write notes with ease in a pop-up window without necessarily opening the Samsung Notes application. To do this, follow the steps below;

- Launch the Air command panel and choose "**Create note.**" Alternatively, tap on the screen twice while you

press and hold the S Pen button. The note screen will show as a pop-up window.
- Start creating a note using your S Pen.
- Click on the **back button** ⟨ when you are done and your note will be stored in **Samsung Note.**

Smart select

Use your S Pen to highlight an area and then carry out some actions, such as saving or sharing. You can also mark a section from a video and then capture it as a GIF animation.

✓ When you see content that you want to capture, like a part of an image, launch the Air command panel and choose **Smart select.**

✓ Choose a shape icon that you want on the toolbar and then drag your S Pen over the content you want to select. If you want to pin the section you selected at the top of your screen, click on ⌖, highlight the area and then choose **pin to screen.**

✓ Choose any of the option below to use with the area that you have selected;
  - **Extract text:** This will allow you to Identify and extract text from the content you have selected.
  - **Pin to screen:** Put a shortcut to the items you have collected on a home screen.

  - **Auto select** : You can allow Smart select to select contents to extract automatically.

  - **Draw** : you can make some drawings on the content you have created.

o **Share** ⌣ : select a sharing method if you want to share your contents with other people.

Unlocking the screen using the S Pen

You can unlock your device's screen if it is locked while the S Pen is still connected by simply pressing the S Pen button. Note that you will only be able to use the S Pen to unlock your screen if you have enabled any of the screen lock methods. Connect your S Pen to your device before you will be able to use your S Pen to unlock your device's screen. To get started, follow the steps below;

- Open the Settings app on the home screen and click on **"Advanced features"**, tap on **"S Pen"** and select **"S Pen unlock."**
- Click on **"Use S Pen unlock"** and click **OK**.
- Unlock your screen with the preset screen lock method.

Now you will be able to unlock your screen by just pressing your S Pen button.

Bixby

Introduction

The Bixby serves as a user interface helping you to use your tablet more conveniently and efficiently. You can use your voice with Bixby or enter text. Bixby will open a function you asked or bring the information you requested. Bixby also learns how you use your device and your environment. The more Bixby gets to learn about you, the more it gets to understand your everyday's actions. Kindly note the following about using Bixby;

- To use Bixby, you must connect your tablet to a Wi-Fi network or any trusted mobile network.
- You need to sign in to your Samsung account before you get to use Bixby.
- Bixby is not available in all languages and some Bixby's features may not work depending on the region of usage.

Starting Bixby

When you open the Bixby as a first user, you will get the Bixby introduction page. Select the language you want to start using with Bixby, sign in to your Samsung's account by following the prompts on the on-screen, and then agree to Bixby's terms and conditions. To get started with Bixby, follow the steps below;

- Press and hold your tablet's Side key. You can alternatively open the Bixby application on the home screen.
- Choose the language you want to use with Bixby.
- Click "**Sign in to Samsung account**" and you will be prompted with instructions to sign in to your Samsung account. If you have already signed in to your Samsung's account, you will see your account information showing up on your screen.

- Follow all the instructions on the screen to finish setting up Bixby. The Bixby screen will show.

Using Bixby

Talking to Bixby through voice or through texts will make Bixby launch the functions you asked for or to bring the information you requested. While pressing and holding on to your tablet's side key, voice out what you want to tell Bixby(what you need Bixby to do for you) and then release your finger when you are done talking. As an alternative, say "Hi, Bixby," and Bixby will give a sound meaning it is ready to hear from you. For instance, hold the side key on your tablet, and ask Bixby "How is the weather today?" and you will see the weather information for that day appearing on your screen. You can still ask about the weather tomorrow while you press and hold on the side key, simply ask "Tomorrow?" and Bixby will bring you the weather information for tomorrow. Bixby is able to do this because it understands that you have asked about weather previously and "Tomorrow" can only mean, "How is the weather tomorrow?"

If Bixby asks any question during the conversation, while still pressing and holding on the tab's Side key, answer Bixby. Alternatively, click on and answer Bixby. If you wear headphones or a Bluetooth audio, or you have already started a conversation with "Hi, Bixby", you will be able to continue your conversation without clicking on the icon.

Open your Bixby application and click on ☰ > ⚙, tap on **Automatic listening** and choose "**Hands-free only.**"

Waking up Bixby with your voice

Although, you can always begin conversation with Bixby by voicing "Hi, Bixby", but for convenience, simply register your voice with Bixby so that Bixby can recognize your voice anytime and it will be able to respond actively after saying "Hi, Bixby." To get started, follow the guides below;

- Open your Bixby application from the home screen and click on ☰ > ⚙ then tap "**Voice wake-up.**"
- Toggle on the **"Wake with "Hi, Bixby"** switch to enable it.
- Complete the registration by following the instructions displayed on the screen.

Now you can voice "Hi, Bixby", and just when your tablet gives a sound, begin a conversation.

Communicating with Bixby by typing text

If Bixby cannot recognize due to one reason or the other – perhaps the surrounding is too noisy or you cannot just speak at the moment – communicating with Bixby through text is better. Open your Bixby application, and select ⬤ and then start by entering your message. Bixby will only respond via text throughout the entire communication and will not use its voice feedback

Bixby usages

Launching apps or services

You can open features and apps very quickly with Bixby. Many convenient services – such as looking for hotels nearby or ordering for coffee – can also be used with your Bixby. For instance, press and then hold on to your Bixby key, voice "Call John," and then lift your fingers from the key. The Bixby app will then send a call to any contact you save as John on your contact list. For convenience, ensure that you have saved only one contact with the name John.

See more ways to use Bixby

To see more ways to use your Bixby, open your Bixby app and then click "**All capsules (or Marketplace**)."

You can look out for services that are supported by Bixby and various examples of commands. Some Bixby's features may not work based on your region or your service provider.

Bixby Vision

Introduction

Bixby Vision is a menu that gives various features based on image recognition. Use Bixby Vision to look for information very quickly by identifying and recognizing objects. Utilize many important Bixby Vision features. Kindly note the following when using Bixby vision;

- To use the Bixby feature, your tablet must have been connected to a good Wi-Fi or a trusted mobile network.
- The number of features and services available for use depend on region or service provider.
- The Bixby Vision feature may not be presently available or you probably may not see correct search information depending on the format of the image, image size, or image resolution.
- The product details given by Bixby are not accounted for by the Samsung Company.

Launching Bixby Vision

You can open the Bixby Vision via any of the method listed below;

- Launch the Camera app from the home screen, click MORE on the list of shooting modes and tap on BIXBY VISION.
- Launch the Gallery app, choose an image and click ◉ .
- From your Internet application, click and hold an image and then tap **Bixby Vision**.
- You can launch the Bixby Vision application from your app screen if you have added it to your app list.

Reading QR codes

Use Bixby Vision to recognize QR codes and access a variety of information, like photos, websites, maps, and business cards.

- Open the Camera application, click on **MORE** on the list of shooting modes, and then choose **BIXBY VISION.**
- Choose and maintain the QR code within your screen to allow Bixby Vision recognize it.

The information associated with the QR code will be displayed.

Searching for shopping information

If there is any product that you want to recognize even without you having any idea of the product's name, you can just use Bixby Vision and your device will search for the product anywhere it is available online, and then bring you the search result.

- Open the Camera application, click on **MORE** on the list of shooting modes, and then choose **BIXBY VISION.**
- Choose and maintain the QR code within your screen to allow Bixby Vision recognize it.
- Tap on a search result to see it.

After choosing a search result, the product information will show.

Searching for similar images and related information

You can search for images that are similar to the recognized object online and some related information about it. You will be able to see various images that have properties similar to the object, like shape or color, and then access the related information.

For instance, if you want to access any information about a car, utilize the Bixby Vision features. The Bixby Vision will scan for and give you related information or relayed images with similar features. To do this,

- Open the Camera application, click on **MORE** on the list of shooting modes, and then choose **BIXBY VISION.**
- Choose and maintain the object within your screen to allow Bixby Vision recognize it.
- Tap on a search result to see it.

After choosing a search result, the similar image and information will show.

Translating or extracting text

Identify and display the translated text on your screen. You can equally extract the text part from any document or an image file.

For instance, if you wish to understand what a sign symbol means while going abroad, deploy the Bixby Vision features on your device. The Bixby Vision feature will decode the text in the sign symbol to any language of your choice. To get started, follow the guide below;

- Open the Camera application, click on **MORE** on the list of shooting modes, and then choose **BIXBY VISION.**
- Select and maintain the text within your screen to allow Bixby Vision recognize it.

The translated text will show on your screen.

- To extract the text, click ⊕. If you click the forward button ❯ , you can utilize some additional features with the text that has been extracted, such as sharing or saving the text.
- Click on the language settings panel located at the bottom of your screen if you want to change the target language or the source.

Bixby Routines

Introduction

Bixby Routines is a menu that automates those settings that you use constantly and repeatedly and then suggests many useful features based on your frequent situations after it has learned your usage patterns and routines. For instance, your 'before bed' routine will carry out some things, like turning on the silent mode and the dark mode, so that they won't be affecting your vision and hearings when you use your tablet before you go to sleep.

Adding routines

You can insert routines to allow you to use your tablet more easily. Set added routines to either auto or manual. The auto setting automatically integrates the routine without necessarily requesting your permission while the manual setting will need you to manually input the routine you want.

Adding recommended routines

When your tablet understands your device usage patterns, the device will recommend you to add some useful features or the features that you frequently deploy as routines.

When you see the recommendation notification, click **View all** and then add the recommendation to your own routine.

Adding routines from the recommendation list

Access the list of useful and important features and then add them as your own routines.

- Open the Settings app and click on "**Advanced features**", and then tap "**Bixby Routines.**"
- On the **Discover** list, choose a routine that you need and then tap Save.
- Click "**Edit**" to assign the conditions and actions for the routine. You can alternatively click on the condition or action that you wish to edit.
    - o If you decide to set your routine's condition to manual, click on "**Start** button tapped."

This selection will show only when no running conditions have been set.

For manual routines, simply add a routine to your Home screen as a widget and then access it as quickly as possible. When you see a pop-up window appearing, click **Add.**

Adding your own routines

Add the features that you wish to use as routines.

- Open the Settings app and click on **"Advanced features"**, tap **"Bixby Routines"** and then choose the **add icon**+
- Tap on the **add icon**+, set the conditions, and then click **Next.** If you wish to set your routine's running condition to manual type, click on **"Start button tapped."**
- Tap on the **add icon**+, set the actions, and then click **"Done."**
- Enter a name for the routine and then click **"Done."**
- For manual routines, simply add a routine to your Home screen as a widget and then access it as quickly as possible. When you see a pop-up window appearing, click **Add.**

Using routines

Running auto routines

The routines you set for automatic will be able to run automatically when their conditions have been detected.

Running manual routines

For the manual routines which you assigned the running condition as **"Start button tapped,"** you will be able to run them manually whenever you tap the button.

- Open the Settings app from the home screen, click on **"Advanced features"**, tap on **Bixby Routines**, select **"My**

routines," and then click next to the routine you wish to run. You can, as an alternative, tap on the routine's widget on your Home screen.

Viewing running routines

Routines that are currently running will show on your tab's notification panel. Click the routine's notification to see any detail about the routine.

Stopping running routines

You can quickly terminate running routines. On your notification panel, click next to the routine and select **Stop.**

CHAPTER FOUR

MORE FUNCTIONS ON SAMSUNG GALAXY TAB S7 AND S7 PLUS

Introduction

Send and answer phone calls and video calls on your device.

Open the **settings app** on your device to start sending and receiving calls and messages. Click on **"Advanced feature"** and then toggle on the **"call & text on other devices"** switch to enable it. You have to register and then sign in to the same Samsung account on your tablet and the other device. You might not be able to access most calling and messaging features.

Making calls

- Open your **Phone** app and then click on **Keypad**.
- Input the phone number that you want to call.
- Click on   to send a voice call, or choose   to send a video call.

Making calls from call logs or contacts list

Launch your **Phone** app, click on "Recent" or "Contacts," and then swipe to your right side on a phone number or a contact to send a call.

If this feature has been deactivated either by default from the manufacturer or by accident on your part, open the **Settings app**, click on "**Advanced features**" select "**Motions and gestures**," and then toggle on the "**Swipe to call or send messages**" switch to enable it.

Using speed dial

Activate the speed dial numbers to make calls quickly.

To assign a particular number to speed dial, open the **Phone** app, click **Keypad** or **Contacts**, tap on the **"more icon ⋮"** and select **Speed dial numbers**, choose a speed dial phone number, and then add a phone number. To send a call, click and hold on a speed dial number on your keypad. For speed dial numbers from 10 and above, select the first digit(s) of the phone number, and then click and hold the last digit of the phone number. For instance, if you assign the number 245 as a speed dial number, tap 2, tap 4, and then tap and hold 5.

Making calls by searching for nearby places

You can conveniently send calls to areas that are near your current location simply by searching for their details. You can search for places nearby by category, like the restaurants or stores, or simply by choosing recommended hot locations.

Open your **Phone** app, select **Places**, and then choose a category or click on the finder Q and then input a business name inside the search bar. Or, choose one from any of the recommended hot places. The business's information, such as its address or phone number, will show.

Making an international call

- Open the **Phone** app and tap on **Keypad**.
- Click and hold on zero until you see the + sign showing up.
- Input the information of the one you want to call, like the country code, phone number and area code, and then click on Q .

Receiving calls

- Answering a call

When you see an incoming call, drag ◯ outside the large circle.

- Rejecting a call

When you see an incoming call that you want to reject, drag ◯ outside the large circle.

If you wish to send a message when you are rejecting an incoming call, push the **Send message** bar upwards and choose a message to send. If you have activated the "Add reminder" switch, you will receive a reminder telling you of the call you have rejected an hour later.

To create various rejection messages, open the **Phone** app, click on the **"more option ⋮"** choose **settings** and select **"Quick decline messages,"** input a message, and then click +.

Missed calls

Any call you missed will show a missed call icon⤵ on your status bar. Swipe down from the notification panel to see the list of calls you have missed. You can alternatively open your **phone** app and select **"recent"** to see the calls you have missed.

Blocking phone numbers

Sometimes, there may be a need to block some specific numbers from calling you, maybe due to spam or any other reasons best known to you. Any contact you blocked will be added to your block list.

- Open the Phone app on your device, click on the **"more option ⋮"** choose **settings**, and select **Block numbers**.
- Click on **"Recent"** or **"Contacts,"** choose the contacts or phone numbers that you want to block, and then click **"Done."**

You can manually input a phone number, click **"Add phone number,"** input a phone number, and then click +.

When any phone number you have blocked tries to send a call across to you, you won't be able to receive the call nor will you get any notification. Their calls will be entering the call log. To block incoming calls from a number that do not show the caller identity, simply toggle on the **"block unknown/hidden numbers"** switch to enable the feature.

Options during calls

- **Add call:** You will be able to dial a second number even while you are on call with the first caller. Your first caller will be on hold so that you can attend to the second call. When the second call has ended, the first call will be put back on.
- **Message:** Enter a message for the caller and send it.
- **Bluetooth:** Change to a Bluetooth headset if you have already connected it to your device.
- **Hold call:** Put a call on hold.
- **Mute:** Turn your microphone off to prevent other parties from hearing what you are talking about on the phone.
- **Keypad / Hide:** Open the keypad or close the keypad.
- Put an end to your current call.

Note that these features may vary based on your region and on your service provider.

Adding a phone number to Contacts

Adding a phone number to Contacts from the keypad

- Open your Phone app and choose **Keypad**.
- Input the phone number that you want to add to your contact.
- Click **"Add to Contacts."**

- Click "**Create new contact**" if you want to make a new contact, or click on "**Update existing contact**" to add the phone number to the list of existing contacts.

Adding a phone number to Contacts from the calls list

- Open the Phone app and click "**Recent.**"
- Select a phone number and click on **Add.**
- Click "**Create new contact**" if you want to make a new contact, or click on "**Update existing contact**" to add the phone number to the list of existing contacts.

Adding a tag to a phone number

You don't necessarily need to save a phone number before you add a tag to it. Adding a tag to a phone number will enable you to access the information about the caller anytime their call comes even when you have not saved their number on your contact list. To add tag to a contact, follow the steps below;

- Open the Phone app and click on "**Recent**"
- Click on a phone number.
- Select "**Add note,**" input a tag, and then click "**Add.**"

When that particular number calls you, you will see the tag under the number.

Contacts

Introduction

Create a new contact or manage existing contacts on your device.

Adding contacts

Creating a new contact

- Open the Contacts application and tap ⊕.
- Choose the location where you want to add the contact to. Note that you can add the contact to your SIM card, memory card (sometimes), Samsung account or your mail.
- Input the details about the contact. Details like contact name (first name, surname and last name), contact email, contact image and some other details.
- Tap on "**Save.**" To save the contact.

Importing contacts

If there is any storage (apart from your device) where you have previously saved your contact, you can add the contacts to your device by importing the contact from wherever you saved them in. If you want to import contacts from your tablet, follow the steps below;

- Open the Contacts app on your device, click on ☰, select **Manage contacts**, choose **Import or export contacts** and then tap "**Import.**"
- Choose a desirable storage location where you want to import contacts from.
- Mark VCF files or contacts to import and click "**Done.**"
- Choose a desirable storage location where you want to import contacts to.

Syncing contacts with your web accounts

Synchronize contacts on your device with online contacts that you have saved in your web accounts, like your Samsung account.

- Navigate to your device's home screen, open your Settings app, click "**Accounts and backup**" choose "Accounts," and then tap the account that you want to synchronize with.
- Click on **Sync account** and toggle the Contacts switch to enable it.

For the Samsung account, click on the "**more option icon** ⋮" choose "**Sync settings**" and click on the Contacts switch to activate the sync settings.

Searching for contacts

Open the contact application on your device and select the Contacts app.

Deploy anyone of the search methods listed below;

- Scroll up or down your list of contacts to find a contact.
- Drag your finger along the index at the right hand side of your contacts list to quickly scroll through your list of contacts.
- You can also search for a particular contact by using the finder ⌕ at the top of the contact list and then enter what you want to search.

Creating groups

You can add groups, such as friends or families, and then manage contacts by group.

- Open your Contacts app and tap on the **"More option icon☰"** select "Groups" and click on "Create group."
- Input the name of the group.

You can set a ringtone for the group by clicking on **ringtone** and selecting a ringtone.

- Click on **"Add member,"** choose the contacts that you want to add to the group, and then click "Done."
- Select "Save."

Sending a group message

You can direct a group message to the group's members at the same time. To do this, open the contact app, click on the **"More option icon☰"**, tap on **"Group"** and then choose a group. Click on ⋮ located at the top of the contact list and select **"send messages"**

Merging duplicate contacts

Importing contacts from any other storage or synchronizing your contacts can create duplicate contacts, which you can mess up your list of contacts. Streamline your list of contacts by merging duplicate contacts. To do this, follow the steps below;

- Open the Contacts app on your device and click on☰ select **"Manage contacts"** and choose **"Merge contacts."**
- Mark all the contacts that you want to merge and click on **"Merge."**

Deleting contacts

- Open your Contacts application, tap⋮ located at the top of your contacts list, and then click **"Delete."**
- Mark all the contacts that you want to delete, and select **"Delete."**

If you want to delete contacts one by one, select a contact and then tap⋮ and choose **"Delete."**

Messages

Introduction

You can send and receive messages on your device with the message feature.

To send out a call and input a text message, open your Settings app, click on "Advanced features," and then select the "Call & text on other devices" switch to enable it. You will need to register and also sign in to the same Samsung account on your device and the other device. There are some calling and messaging features that may otherwise not be available.

Sending messages

During roaming, additional charges may be applicable when you send out a message. To send a message, follow the steps below;

- Open your Messages application and tap $+$.
- Add the number or the contact that you want to send the message to, and then enter your message.

You can record and send a voice message by tapping and holding on the , say the message and lift your fingers. The recording icon will only show while the message entry field has not been filled with any text.

- Click on the to send your message.

Viewing messages

All of the incoming messages are usually arranged into threads according to contacts. During roaming, additional charges might be applicable when you receive a message.

- Open the Messages application.
- Tap on a contact or a phone number.
- Click on the entry field to reply to a message, enter your message, and then tap on .
- By pinching on your screen or spreading your two fingers apart on the screen.

Sorting messages

Sorting messages allow you to manage messages according to category. To sort messages, follow the steps below;

- Open the Messages app on the home screen and click on **Conversations**.
- Click on "**New category**" and select "**Add category**." If you cannot see the category option, simply tap on ⋮, select **Settings** and then toggle on the "Conversations categories" switch to enable it.
- Input a name for the category and click "**Done.**"
- Choose conversations that you want to add to the category and then click "Done."

Blocking unwanted messages

If you block a number from sending you messages, they won't be able to reach you via message. The number will be added to your block list. To block any unwanted number, follow the steps below;

- Open the Messages app on your device; select the "**more option icon** ⋮ located at the top section of the messages list, click on **Settings**, select **Block numbers and messages** and then click on "Block numbers."

- Click on **Conversations** and select a phone number or contact that you want to block. Alternatively, click on **contact,** select contact, and then choose "Done." You can manually enter a phone number under "**enter phone number**" and choose the + icon.

Setting the message notification

You can change your display options, notification sound and a lot much more. To do this, follow the steps below;

- Open the message application, click on the "**more option icon** located at the top section of the message list, click on **settings > notifications** and then toggle on the switch to enable notification.

- Change the message notification settings.

Setting a message reminder

You can configure an alert at regular intervals to remind you that you have some unchecked notifications. To enable this feature, open the **settings app,** click on **"accessibility",** tap on **"Advanced settings >notification reminder"** and toggle on the switch to enable it.

Deleting messages

- Open the message application on your device.

- Choose a contact or a phone number on your message list.

- Tap on a message and hold, then select "Delete."

To delete many messages at a time, mark messages that you want to delete.

- Select "Delete."

Samsung DeX

Samsung DeX is a platform that can enable users to use their tablet like a computer just by connecting the tablet device to an external display, like a TV or monitor, or to the Keyboard Cover. You can bring tasks from your tablet device to your large screen (TV or monitor) with the aid of a keyboard and mouse. You will still be able to use your tablet device while using the Samsung DeX platform.

Connecting devices and starting Samsung DeX

Connecting to the Keyboard Cover

You can achieve a PC experience on your tablet by connecting a keyboard cover to your tablet. The keyboard cover has a connector that can fit right in the tablet's keyboard dock. Once the keyboard dock of the tablet has fit in the keyboard's cover connector, you will see the Samsung DeX screen coming up on your tablet.

- Set your tablet to automatically switch to Samsung DeX anytime the Keyboard cover has been connected, open the **setting app,** click on **"advanced features"**, select

"Samsung DeX" and toggle on the **"Auto start when Book Cover Keyboard is connected"** switch to enable it.

- When you have successfully connected the Keyboard cover to your tablet, you can open the Samsung DeX using a key combination by tapping on the **Fn + DeX.**
- You can also open the Samsung DeX platform right from quick settings without the Keyboard Cover. To do this, swipe downwards on your notification panel and click on "DeX."
- To use your tablet in the upright position, join the backside of your Keyboard Cover to the back of your tablet.

Wired connection to an external display

Connect your device to an external display by using an HDMI adaptor.

- Connect the HDMI adaptor to your device.
- Attach an HDMI cable to the HDMI adaptor and also to a monitor's or a TV HDMI port.
- On your device's screen, click on **Start**.

Without changing your device's screen, the Samsung DeX screen will show on the TV or monitor you have connected.

- Set your tablet to automatically switch to Samsung DeX anytime the HDMI Adaptor has been connected, open the **setting app,** click on **"advanced features"**, select **"Samsung DeX"** and toggle on the **"Auto start when Book Cover Keyboard is connected"** switch to enable it.

Connecting to a TV wirelessly

Using Samsung DeX while connecting to your TV wirelessly is an amazing experience you can always try out. To do this, follow the steps below;

- On your device, launch your notification panel, swipe downwards on the panel, and then tap DeX and hold on it.
- Choose your TV from the list of TVs that has been and click on **"Start now"**
- You will be prompted to honor the connection request on your TV.
- Complete the connection by following all the instructions displayed on the screen.
- When the two devices have been connected, you will see the Samsung DeX screen appearing on your TV.
- The Samsung Smart TV made by Samsung after 2019 should be used with DeX on your device as it is the most recommended.
- You should ensure that the TV you want to connect to your Samsung DeX supports screen mirroring.

Controlling the Samsung DeX screen

Controlling with Keyboard Cover's touchpad and keyboard

When you deploy the Keyboard cover with the Samsung DeX, you can use your Keyboard Cover's touchpad and keyboard to easily control the Samsung DeX screen.

Controlling with an external keyboard and mouse

Use a wireless keyboard/mouse.

- You can navigate with the mouse pointer in order to flow from your external display to the screen of your tablet. To do this, Open the Settings app, choose **Samsung DeX**, tap on "**Mouse/trackpad**," and then choose the "Flow pointer to tablet screen" switch to enable it.
- You can as well deploy the external keyboard on your tablet's screen.

Using your tablet as a touchpad

You can deploy your tablet as a touchpad and work around it with your fingers or your S Pen.

- On your device, swipe downward from the top section of your screen to launch the notification panel and click "**Use your tablet as a touchpad.**"
- Once you launch the notification panel, you can see the gestures to use with the touchpad.
- If your device's screen turns off automatically or by accident, press the Side key or tap twice on the screen to turn it back on.

CHAPTER FIVE

BIOMETRIC AND SECURITY

Face Recognition

- If you are using your face to lock your screen, you won't be able to unlock your screen with your face the first time you are turning on your device. To unlock your tablet, you need to unlock your screen using the PIN, password or the pattern that you set during face registration. This is why you need to be conscious not to forget your PIN, pattern or password.
- The moment you change your screen lock to "**None** or **swipe**" method (these methods are not secure methods to lock your device), all of your biometric security data will be lost. If you need to use your biometric data in applications or features, you have to register your data on your device again.

Tips for using face recognition

Before you register your face as a means of unlocking your device, you need to understand the following information;

- Your tablet could be unlocked by anyone or anything that looks like your picture.
- Face recognition is not as secured as using PIN, pattern or password.

For better face recognition

Consider the following when using face recognition:

- Endeavor to remove your glasses, beards, hast, masks or heavy makeup while registering your faces. These things can disturb the process. And if you can't remove these things, you need to tell during face registration so that your device can understand the conditions with which you register.
- Ensure that you are in an environment that is well illuminated and that your camera lens is thoroughly cleaned when you want to register.
- Don't use a blur image for a good match result.

Registering your face

To have a better face registration experience, kindly register your face in your house or room and stay away from direct rays of the sun. To register your face(s), follow the steps below;

- Navigate to the **settings** app on your device, click on **"Biometrics and security"** and select **"Face recognition."**
- You will be prompted with some instructions on the screen, read and understand these instructions and click **continue.**
- Assign your preferred screen lock method.
- Choose whether you are putting on glasses or not and click **"Continue."**

- Hold your tablet with the screen facing you and put your face and eye on the screen.
- Place your face in the frame on your screen. The camera will start scanning your face to recognize it.
- If unlocking your tablet screen using your face is not working properly, choose "**Remove face data**" to delete your face that has already been registered and then register your face again.
- You can enhance the face recognition by adding an alternate appearance, simply choose "**Add alternative look**"

Deleting your registered face data

To delete the face data that you have already registered, follow the steps below;

✓ Navigate to the **settings app** on your device, click on "**biometric and security**" and choose **"Face recognition."**
✓ Unlock your device's screen by using the preset screen lock method.
✓ Click on "**Remove face data**" and choose "**Remove."**
Once your registered face has been deleted successfully, all the features related to your face data will also be removed.

Unlocking the screen with your face

You can unlock your device's screen with your face without using a PIN, pattern, or password. To do this, follow the steps below;

✓ Navigate to the **settings app** on your device, click on "**biometric and security**" and choose **"Face recognition."**

✓ Unlock your device's screen by using the preset screen lock method.

✓ Click on the **"Face unlock"** switch to enable it.

- If you want to set your tablet to unlock your screen without swiping on the locked screen once it has recognized your face, choose the "Stay on Lock screen" switch to deactivate it.

- If you want to limit the possibility of recognizing just about any faces that look like you in photos or videos, select the **"Faster recognition"** switch to deactivate it. This may reduce the face recognition speed.

- If you want to set your tablet to recognize your face only when you open your eyes, choose the "**Require open eyes**" switch to enable it.

- If you wish to increase the rate of face recognition in a dark environment, choose the "Brighten screen" switch to enable it.

✓ Look at your screen on the locked screen.

When your face has been recognized, you can get to unlock your screen without deploying any additional lock screen method. If your face is not recognized, utilize the preset screen lock method.

Fingerprint Recognition

Registering fingerprints

- Navigate to the **settings** app on your device, click on **"Biometrics and security"** and select **"Face recognition."**

- You will be prompted with some instructions on the screen, read and understand these instructions and click **continue.**
- Set your preferred screen lock method.
- Register your fingerprint by following the steps below;
    - o **For some Tab S7 Variant:** Put your finger on the device's side key. Remove the finger once your device has detected the finger and put the finger on the side key once again.
    - o **For other Tab S7 variants:** Put your finger on the device's fingerprint recognition sensor. Remove the finger once your device has detected the finger and put the finger on the fingerprint sensor once again.
    - o Repeat these actions until your fingerprint has been duly registered.
- Click "**Done**" after you are done registering your fingers.

Deleting registered fingerprints

To delete your registered fingers, follow the steps below;

✓ Navigate to the **settings app** on your device, click on "**biometric and security**" and choose **"Face recognition."**
✓ Unlock your device's screen by using the preset screen lock method.
✓ Choose a fingerprint that you want to delete and then click "**Remove.**"

Unlocking the screen with your fingerprints

Unlock your device's screen with your fingerprint rather than using PIN, pattern or password.

✓ Navigate to the **settings app** on your device, click on "**biometric and security**" and choose **"Face recognition."**
✓ Unlock your device's screen by using the preset screen lock method.
✓ Click on the "**Fingerprint unlock**" switch to enable it.
✓ On your locked screen, put your finger on your fingerprint recognition sensor to scan your fingerprint.

Wi-Fi

Connect to the internet over a Wi-Fi network by activating the Wi-Fi feature on your device.

Connecting to a Wi-Fi network

- From your Settings screen, choose **Connections**, select **Wi-Fi** and toggle the switch to enable it.
- Choose a network from the list of Wi-Fi networks available. If the network that you want to connect to needs a password, you will see the network having a lock icon. Simply enter the network password to connect to the network.
- Once your tablet connects to the Wi-Fi network, the tab will be able to reconnect to that network another time it is available without asking for a password. To prevent your device from automatically connecting to that network, click on the **settings icon** ⚙ next to the network and click on the "**Auto reconnect**" switch to disable auto reconnection.

Bluetooth

Pairing with other Bluetooth devices

- On your Settings screen, select **Connections**, choose **Bluetooth** and click on the switch to enable it. Your Bluetooth will scan for a list of available Bluetooth devices.
- Choose a device that you want to pair with. If the device that you want to pair with is not available on the list, set that device to enter Bluetooth pairing mode. Your device will be seen by other devices while your Bluetooth settings screen is open.
- A Bluetooth connection request will be sent to your device, honor the connection to confirm. Both devices will be connected when you have accepted the connection request.

Sending and receiving data via Bluetooth

Many applications support transferring data via Bluetooth. You can share information, such as media files or contacts, with other Bluetooth devices. Follow the steps below to get started,

- Open your Gallery app to select the image that you want to send.
- Click on ⌁, choose **Bluetooth** and choose the device that you want to transfer the image to. If that device that you want to pair with is not available on the list, ask the device owner to turn the device visibility option.
- Ask the other device to accept the connection request.

Unpairing Bluetooth devices

❖ From your Settings screen, choose **Connections**, and click on **Bluetooth**. The device will bring the paired devices in the list.
❖ Click on ✿ next to the name of the device to unpair.
❖ Select "**Unpair.**"

CHAPTER SIX

CAMERA APP

Launch the camera application to take and edit pictures as you wish.

Launching Camera

Deploy the following means to open your Camera application:

- Navigate to the home screen and open the **camera app.**
- Press the Side key two times, albeit quickly.
- Drag the camera icon outside the circle from the lock screen.
  - o Depending on your region or your service provider, some methods may not work.
  - o Some camera features may not be available when you open your Camera app from your locked screen or when your device's screen is turned off while the screen lock method is activated.
  - o If your camera app keeps taking blurry images, clean the lens of the camera and take the picture again.

Taking photos

- ❖ Click on the image on your preview screen where your camera should focus.
  - o Spread your two fingers apart on your screen to zoom in, and pinch using your fingers to zoom out. You can alternatively drag the camera lens selection icon. You

can only use the camera zooming features with the rear camera.

- o Tap on the screen if you want to adjust the photo's brightness. When you see the adjustment bar, drag the adjustment bar towards the"**+**"to increase the brightness or towards the"**—**" to reduce the brightness.

❖ Tap on the icon at the center⭕ to take your photo.

Using the camera button

Click and hold on your camera button if you want to record a video.

Take a burst shot by swiping your camera button to the screen and hold it.

Options for current shooting mode

Utilize the following options on the preview screen.

⚡ Tap on this icon to activate or deactivate your camera's flash.

- ⏱ : Choose to what length your camera should be delayed before it starts taking photos automatically.

- [4:3] : Choose an aspect ratio for your photos.

- ▣ : Set this to let your device camera take a video clip for some seconds before you tap on ⭕.

- ☾ : Tap on this to enable or disable your device's night hyperlapse feature.

- : Choose a frame rate.

- (16:9) : Choose aspect ratio for your videos.

- : Apply beauty filter or filter effect to videos.

Choosing a camera for shooting

Choose a camera that you want on the preview screen and then start recording your video or take a picture. You will only be able to deploy this feature in some shooting modes as it is not available in all shooting modes.

- **(Wide-angle):** The wide-angle camera enables users to record normal video or snap basic pictures.

- **(Ultra wide):** The Ultra wide camera allows you to snap wide-angle photos or to record wide-angle videos that look like the actual view. Utilize this feature to snap landscape pictures.

To correct any distortion in photos you took using the Ultra wide camera, click on your preview screen, select "**Save options,**" and then choose the "**Ultra wide lens correction**" switch to enable it.

Taking selfies

Use your device's front camera to snap self-portraits. You can take your portraits by yourself or with your friends. To do this, follow the steps below;

- ✓ Tap on PHOTO on the shooting modes list.
- ✓ On your preview screen, you can swipe upwards or downwards to switch camera, or choose ⊙ to change to the front camera that you can use for taking self-portraits.
- ✓ Put your face in the front camera lens. To snap self-portraits using a wide-angle shot of the landscape or people, click ⊞ .
- ✓ Click on ◯ to snap a photo.

Video mode

Use your device camera to shoot video. Your camera is able to adjust the mode of shooting automatically to record videos. To record video, follow the steps below;

- Select VIDEO on the list of shooting modes.
- Tap on ◉ to start shooting your video.
- Switch between the rear and the front camera while recording by either swiping upward or swiping downward on the preview screen or click on ⊙
- To take an image from the video during recording, click ◉ .
- Select ▣ to terminate video recording.

About the Author

Konrad Christopher is a video software expert with several years of experience in videography and software development. He is consistent following the latest development in the tech and software industries and has an eye for high-end video equipment and software. He loves solving problems and he's enthusiastic about the software market.

Konrad holds a Bachelor's and MSc degree in software engineering from Cornell University, Ithaca. He lives in New York, USA. He is happily married with a kid.